Deserts are found on every continent except Europe and cover about 20 percent of Earth's land surface. Deserts are dry and harsh, most receiving less than 10 inches of rain a year. Yet many plants and animals live there. One desert with many different kinds of plant and animal life is the Sonoran Desert. It stretches north from Mexico into Arizona and southeastern California.

The Saguaro (sah-WAH-roh) cactuses that dot the Sonoran Desert will bloom after sunset in May or June. Can you find the buds, flowers, and dying flowers?

Later, the Saguaro's flowers fall off and rich red fruit grows.

The Five Seasons

The Sonoran Desert is unusual because it has five seasons. The rainy seasons are late summer and winter. The dry seasons are spring, early summer, and fall. Since the desert is dry much of the year, most desert animals and plants have to get by with very little water.

Many desert animals, like the Antelope Ground Squirrel, get the water they need from food, but will drink from canyon streams or water holes, when they have the chance.

Temperatures may soar to 120 degrees Fahrenheit in the early summer (May and June). The desert is very dry then, as it is in the fall and spring, too.

Rain may pour down in the summer rainy season (July through mid-September).

During the rainy season, rain may be so heavy that a dry streambed, called an *arroyo* (ah-ROY-yo), can fill up with water in less than an hour.

Snow in a desert? There may be in the winter (December through mid-February)—the Sonoran Desert's other wet or rainy season.

Amazing Plants

In the Sonoran Desert, plants come in more shapes and sizes than in any other desert in North America. But how do they survive in such a hot, dry place? Some, like cactuses, store water. Because much water is lost from plant leaves, many desert plants drop their leaves to save moisture.

If winter has been rainy, spring flowers may carpet the desert sand for miles. After a few weeks, the plants wither, leaving new seeds that may begin to grow again after another rainy year.

Like other cactuses, the Organ Pipe stores water in its waxy, accordionlike trunks and branches, which expand to hold water.

The leaves of the Creosote (KREE-oh-sote) bush have a waterproof coating that helps prevent the plant from losing water. If it stays dry for a long time, the plant will drop its leaves and lose branches. But, shortly after rain, the Creosote grows new ones.

After rain, this thorny desert plant, a Fire Thorn or Ocotillo (oh-kuh-TEE-yo), grows leaves. When dry, the plant drops its leaves to save moisture. Even if the soil is dry, its pretty red flowers will bloom in the spring.

A type of Prickly Pear, this cactus' stems grow at certain angles to protect them from the drying rays of the sun.

The Fishhook Barrel cactus has curved fishhook spines that blunt the wind and provide shade.

Much of the year, the Paloverde (PAH-loh-VEHR-day) or "green stick" tree has no leaves, and its branches and twigs make its food. Can you see its green trunk and branches?

Remarkable Animals

Like desert plants, the animals have made amazing adaptations. Over millions of years, animals have adapted their bodies and altered their habits to beat the heat or to find food, shelter, and protection from predators. Some go into long periods of *dormancy,* or a resting state. Others are active only at night when it's cooler. Many have found ways to live with less water and food.

The blood vessels in the Antelope Jackrabbit's large ears give off heat from its body and help keep it cool.

Strong legs help make the Kangaroo Rat an excellent burrower that digs deep to keep cool and safe. This rodent never needs to drink water even though it eats dry seeds.

The Desert Spadefoot Toad may stay buried underground for as long as 10 months after eating only a single meal! The toad surfaces after a heavy rain, just long enough to find a mate and lay eggs.

One of only two poisonous lizards in the world, the Gila (HEE-lah) Monster can go a year or more without eating, living off the fat stored in its tail.

The Roadrunner seldom flies, but runs at great speed—up to 15 miles an hour. Its four long toes—two pointing forward and two pointing back—allow it to dash after even the fastest lizard. This desert bird gets its water from the food it eats.

Desert Defenses

Desert animals must find food to live, but sometimes the hunters may become the hunted. Clever actions, speed, and even unpleasant smells help keep some desert animals safe from predators.

If a predator approaches, the Chuckwalla (CHUK-wah-lah) dashes to safety among rocks. Inflating its body, the lizard easily wedges itself into a narrow crack.

The Killdeer makes its nest in the open. If its young are in danger, the bird will pretend it has a broken wing to lure the predator away from its nest.

Few predators will get near the Badger because it gives off a skunklike smell and can be a fierce fighter.

The Pronghorn can outrun most of its predators—at speeds of more than 50 miles an hour!

When caught, the Regal Horned Lizard may shoot blood from its lower eyelid as far as 4 feet—so distasteful is this to a predator that it may choose another meal!

Keeping Cool

Many desert animals are *nocturnal,* only active at night. Some may be active early or late in the day, avoiding the extreme midday heat.

The piglike Collared Peccary (KAH-lerd PEH-kah-ree), or Javelina (ha-veh-LEE-nah), is often active in the early morning, as well as at night.

The Kit Fox is mostly nocturnal, doing its hunting at night. During the day, it seeks shelter from the heat in an underground burrow or den.

Like many desert animals, the Ringtail sleeps by day and hunts at night.

Desert Look-alikes

Appearances may be deceiving. Some desert creatures blend in so well with their surroundings that they may be difficult for other animals to spot. This is called *camouflage* (KA-muh-flaj). Often, it might be hard to tell if they are coming or going!

Well hidden in the Creosote bush, the Coachwhip Snake waits to ambush its next meal.

The leaves of the Creosote hide this Creosote Bush Katydid, certain grasshoppers, and other insects that live only on this bush.

The Giant Centipede (SEN-tuh-peed) has a painful bite. To avoid being bitten, a predator should attack the head first, but the predator may have a problem since the Centipede's head looks like its tail end.

This Northern Walkingstick is on a tree. It would be hard to see on a desert bush!

Saguaro Cactus Home

Called the "skyscraper of the desert," the Saguaro cactus shows how important desert plants and animals are to each other. Birds, bats, and insects visit, feed from, or nest in the Saguaro. But they also help the plant since they carry pollen from flower to flower, helping the cactus produce fruit and seeds.

The hearty Saguaro cactus may grow to be more than 40 feet tall and live to be 200 years old. It can take 75 years to begin to grow an arm. Although few Saguaros do, this one sports a crown, or fan-shaped top. What causes this is unknown, but it could be from lightning or damage from the cold.

The Gila Woodpecker chisels out its nest in a Saguaro and also feeds the plant's fruit to its young. The temperature inside the cactus is 10 to 20 degrees cooler than the temperature outside during the day, and 10 to 20 degrees warmer at night—keeping the young birds comfortable.

"Who cooks for you-all?" coos the White-winged Dove from its nest on top of the Saguaro. From here, it can dip down for its next meal—the Saguaro's red, figlike fruit.

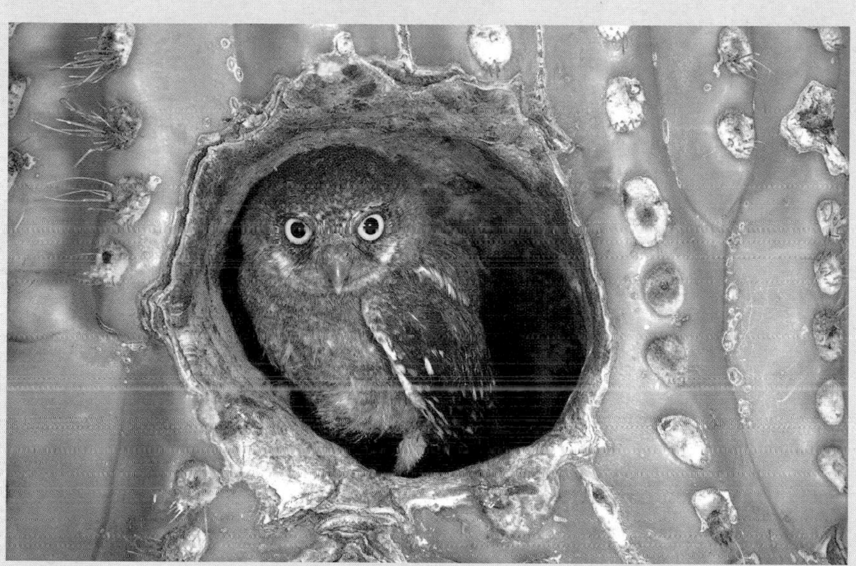

When the Woodpecker moves out, the tiny Elf Owl may move in. A nocturnal desert bird, the Elf Owl feeds on insects that it finds in or near the Saguaro.

A Cougar is poking around the Saguaro. It eats large mammals primarily, but may eat mice, birds, and even grasshoppers.

Honeybees find a Saguaro skeleton a good place to build a hive.

Woody ribs support the Saguaro, which may grow to weigh several tons! After the plant dies, the remaining skeleton becomes home to birds like this Inca Dove, its coloring closely blending in with its surroundings.

Desert Rest Stops

Other cactuses are useful to desert animals, too. Cactuses provide food and places to perch or lay eggs. The base of a cactus makes a good home for some, while spines can help make homes safe.

Cholla (CHOY-ya) cactus spines don't discourage baby Antelope Ground Squirrels.

The Desert Tortoise eats Prickly Pear fruit, spines and all.

The White-throated Woodrat uses pieces of spiny Cholla cactus to cover the entrance to its home. These help to discourage predators. Look closely and you'll see that the Woodrat has also carried home some food—the red fruit from a Saguaro.

The Prickly Pear fruit is almost as big as this Cactus Mouse, which may make its nest among clumps of cactus.

Do you see the tiny white eggs on these Barrel cactus spines? These are fly eggs!

People, too, have found a home in the desert. But, as their communities expand, the desert habitat—also home to the Harris Hawk and so many other animals and plants—will grow smaller.

People need to be more aware of the effects of their activities on the desert habitat so that it can be preserved for all species. In this way, people, animals, and plants can continue to share their desert home.

of the same characters. The stories in this book, however, take place two generations before the events that begin in *Dawn Land* in which people such as Rabbit Stick and Sweetgrass Girl are grandparents.

Whenever I write down the stories from this long-ago time I always hear a voice speaking those tales. It is not my own voice but the voice of a wise elder who is honored for his ability to speak and remember. Join me now in listening to him.

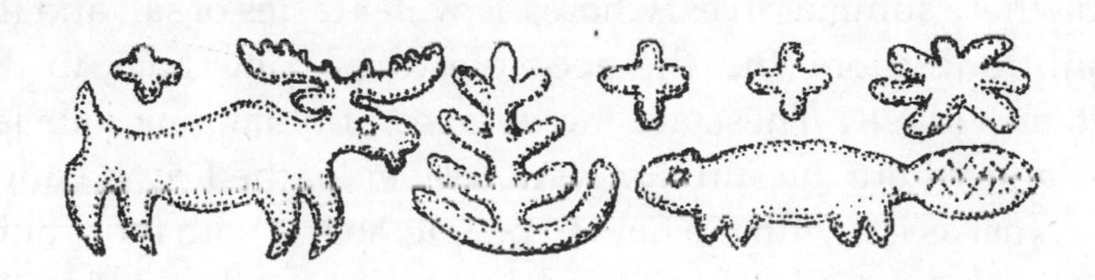

Hear my stories. This is the place where my stories camp. These are good stories. They are stories about the Human People and the Dog People. They are stories about the young ones.

There are many good places between the wide lake the people call Petonbok, the Waters in Between, and Kwanitewk, Long River. That is why the thirteen villages are there. Each one is in a place like no other. At one place the berry bushes are thick with fruit. At another the salmon come up the long river in such numbers that people with fish spears can catch twice their own weight in fish in the time it takes the Day Traveler to move a hand's width across the sky. In another place there are many birch trees which can give up their blankets of bark to make wigwams that will keep out the winter's cold breath. In this land the deer and the moose, the caribou and the great horned elk can be found in big herds. The people who remember to respect the game animals are always successful when they hunt.

Those places, each within their own watershed, their own little valley with good springs to drink from and dry land on which to build homes, are so good that the people always come back to them. The people move their camps to the waters when the fish run or to the hills when the berries are ripe. They travel often to visit each other's villages. They go as far as the Aways Winter Land where the great mountains of ice still live or the land where summer lives, where the water tastes of salt and there is only one shore. But the people always come back to these thirteen villages. These are the villages of Ndakinna, our land. Our land where the sun first rises, Wabanki, the Dawn Land.

Just as it is with the human people, so it is with those animal people who chose long ago to make our lodges their lodges, too. Even before the great mountains of ice traveled back to the always winter land, the Dog People were with us. They stay with us. Although they have always been free to return to the forests where their older brothers, the wolves, still remain, they have chosen to be by our sides.

Ktsi Nwaskw, the Great Mystery who made all, is very wise and wants us all to appreciate each other. So, just as every day has been made to be different from every other, Ktsi Nwaskw made every human person and every animal person to be unique. Each has been given their own story, and from the time they are very young, their stories continue to grow and change. So I will tell you of six young human people and six young dog people. I will tell you of Hummingbird and her dog Awasosis, Little Bear. I will speak of Cedar Girl and her dog Azeban, Raccoon. I will speak of Muskrat and his dog Kwaniwibid, Long Tooth. I will tell of Keeps-Following-the-Trail and his dog Soksemo, Good Nose. I will speak of Sweetgrass Girl and her dog Moosis, Little Moose. I will tell of Rabbit Stick and his dog Mikwe, Squirrel.

I am Ktsindatlogit, Great Storyteller. So, I will speak truly. But you must listen. Are you listening? Unh-honh!

Now hear.

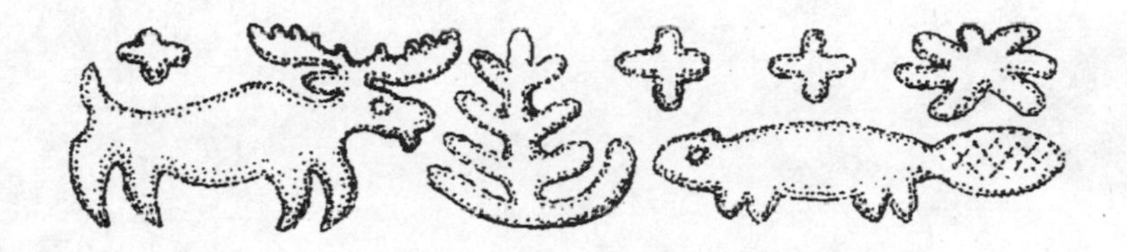

HOW THE DOGS BECAME COMPANIONS TO THE PEOPLE

HUMMINGBIRD MEETS AWASOSIS

The bright warmth from the Day Traveler felt good on the little girl's face as she came out of the wigwam and looked up. But she did not stand there long. This was the morning she had been waiting for. Although there were slippery patches of snow melting at last back into the earth in this Moon When the Winds Are Strong Blowing, the feet of Nanatasis were sure as she darted across the small clearing, moving as swiftly as the little quick flyer which had given her its name—Hummingbird—by coming right into their lodge on the day of her birth.

Hummingbird slowed her steps to a soft, careful walk when she was within a few paces of the little lean-to which barely was as high as a man's waist. She knew that loud sounds or quick motions would be disturbing.

"Awasosqua," she called softly, her voice almost as sweet as the call of one of the little bright-colored singers which would soon flock in great number in the treetops.

From inside the lean-to, the growling sound turned into a gentle whimper.

Hummingbird bent slowly and looked into the lean-to. Awasosqua lay there on her side. She looked at Hummingbird in a way that told the small girl she understood what this day was. Only one of her six puppies was feeding. The others lay sleeping. Some were curled into little balls of fur, others were spread out trustingly on their backs. The smell of the puppies was so sweet and they looked so soft, as soft as the fur of a rabbit, that Hummingbird wanted to crawl in among them and close her own eyes.

But Hummingbird remembered what day this was and how she had been told by her mother to behave. She could not act like a baby if she was going to take on the role of a mother.

"Awasosqua," Hummingbird said, "you are a good mother."

Awasosqua whimpered again and wagged her tail. It was clear to Hummingbird that the big dog understood every word she said.

"Now," Hummingbird continued, "the time has come for me to help you. I will be the one to care for one of your children. She has told me that she wishes to come and live with us in our lodge. Since it is your lodge also, you will still be close to her. But she has told me that she wishes me to be her great friend, to have adventures with me—just as you have done with my mother. So, now that she is old enough, I have come to bring her to our lodge."

Awasosqua did not whimper. Instead, she looked deeply into the little girl's eyes and then bent her head to nuzzle at the fat puppy that lifted up its head, eyes still fuzzy with sleep as it yawned.

"Onnh-honnh," Hummingbird said. "She is the one."

Hummingbird held out her thank-you offering to the big dog, a tough piece of dried moose meat almost as big as one of the puppies. Awasosqua opened her mouth and carefully took it from the little girl. She did not begin to chew it. Instead, she placed it on the floor of the lean-to, where hemlock boughs had been woven together by Hummingbird's mother to make a comfortable, clean bed for the big dog and her puppies. Her eyes found the eyes of Hummingbird and the little girl understood. She leaned over and grasped the fat puppy under its front legs. The puppy was so heavy that she had a hard time lifting her and Hummingbird almost lost her balance and fell back. But her responsibility made her stronger and she did not fall.

"Awasosis," Hummingbird said, "Little Bear, you are going to live with me as my friend now."

The puppy did not whimper as Hummingbird lifted her. She opened her mouth to yawn again and then began to lick the little girl's face.

That night, as the family sat around the fire in the lodge, Hummingbird turned to her mother, Sings Like a Thrush. Awasosis lay peacefully by the little girl's side, her head in her lap as Hummingbird continued to stroke her long, silky ears.

"Mother," Hummingbird said, "why is it that the dog people live with us? Why is it so easy for them to leave their own families and choose to come into our lodges and be our friends?"

"Long ago," her mother said, "in the time before there were any human beings, Gluskabe, The One Who Does Much Talking, the one who helps our Creator, that one was out walking around."

Hummingbird leaned forward. This was her favorite time, the time when stories would be told. She smiled as she noticed that her older brother, Nolka, who was leaning against their father, working with a piece of deer antler to sharpen a spear point by pressing off flakes of stone at its edges, shifted just slightly toward the fire so that he could hear better, too. Even though this story was being told for his little sister, he was not going to miss hearing it, either. Like the deer which had given him his name, Nolka's ears would always be alert to hear any new sound—especially if that sound was their mother's voice telling one of the ancient tales.

"'Soon,' Gluskabe said, 'it will be time to make the human beings. I will be glad to see them. But I wonder how the animal people will treat them?'

"So Gluskabe decided to call the animal people together to ask them how they would treat these new ones who were about to be created. He went into a clearing in the forest and called them. 'All you Animal People,' he called, 'come here to me!'

"So, from the tiniest beings to the biggest ones, they assembled. From the littlest mouse to the giant bear, the Animal People gathered around Gluskabe. See them there, waiting for Gluskabe to speak!"

Their mother paused and looked around the wigwam. Everyone was listening. Even their father had put down the shaft he had been readying to hold the spearpoint that Nolka was shaping. Sings Like a Thrush was the kind of storyteller whose words made your feet walk those paths she described. It was as if all those animals of long-ago had gathered with them inside the little round lodge.

The puppy squirmed and then whimpered, breaking the dramatic silence. Hummingbird buried her face in the soft fur on the puppy's neck and whispered to her. "Listen well, Awasosis, this story is being told for you."

Sings Like a Thrush did not seem to hear. Her eyes were far away, seeing the story as she began again to tell it.

"'Kita, nidobak,' Gluskabe said. 'Listen, all my friends. Soon new ones are going to be created and I want to know how you will treat them. I want each of you to come close. Then I will whisper the name of the new ones who are going to be created and you can tell me what you will do.'"

"This is a good part," Hummingbird whispered, "listen, Awasosis!"

Nolka raised an eyebrow and looked at her across the fire. He was almost losing patience with her interruptions. But he knew that older brothers had to put up with such things, and he simply sighed and leaned back against his father.

A very small smile passed across the face of Sings Like a Thrush. Then she raised her cupped hands to either side of her head to imitate the ears of a bear.

"Great Bear was the first to come forward. He was very proud of himself because he was the biggest of the animals, much bigger than bears are today. His teeth were as long as my arm and each of his feet were as big as a wigwam! Gluskabe came close to Great Bear and whispered the word for human being into his ear. 'Alnobak,' Gluskabe whispered.

"Great Bear reared up on his hind legs and roared. He swung his paws in the air. 'I will tear them apart!' he growled. 'I will swallow them whole!'

"Gluskabe put out his hand and picked up Great Bear. 'Nda,' Gluskabe said. 'You are too fierce.' Then Gluskabe gently stroked Great Bear with his right hand. Great Bear became smaller and smaller until he was not much larger than bears are today.

"'Now you will not be such a danger to the human beings,' Gluskabe said. He put Great Bear down. 'Run away into the forest,' Gluskabe said and Great Bear ran away.

"Now Great Moose came forward. In those ancient days, Moose was taller than the tree tops with sharp horns bigger than the branches of the giant pine. 'Alnobak,' Gluskabe whispered into the ear of Great Moose.

"Great Moose stomped his feet. 'I will throw them up into the air with my horns,' he bellowed. 'I will trample them under my hooves.'

"Gluskabe reached out his hand and picked up Great Moose. 'Nda,' Gluskabe said, 'that will not do.'

"Then Gluskabe pushed on the nose of Great Moose with his right hand. As he did so, Great Moose grew smaller and his nose was pushed in —just as it is today. Gluskabe pressed down on the horns of Great Moose and they flattened out and were no longer so sharp and dangerous. 'Now you will not be such a danger to the human beings,' Gluskabe said. 'Run away into the forest.'

"The next to come forward was Great Squirrel. In those ancient days, Great Squirrel was the fiercest of all the Animal People. Great Squirrel was almost as large as Great Bear and was very strong. 'Alnobak,' Gluskabe whispered.

"Great Squirrel jumped up and down and tore huge branches off the trees, throwing them down onto the ground. 'I will chew them up,' Great Squirrel shouted in a terrible voice, 'I will throw trees down on top of them and crush them.'

"Again Gluskabe reached out his hand. He picked up Great Squirrel and began to stroke him. But because Great Squirrel was so very angry and fierce, he stroked him longer than Great Bear or Great Moose. When he finished, Great Squirrel was great no longer. He was smaller than the rabbit. 'You are too fierce,' Gluskabe said.

'It would not be safe for you to remain as a big animal. Now run away into the forest.'

"Squirrel ran away and climbed up into the tree tops. But although Squirrel was now very small, Squirrel was still angry and fierce. To this day, sometimes when you walk under a tree Squirrel will shout at you and throw down branches to try to crush you. But because Squirrel is small, Squirrel's voice is small, too, and the branches Squirrel throws are no bigger than little twigs."

Hummingbird laughed at the thought of Squirrel trying to crush people with little twigs. Her father and her brother were smiling, too.

"Now the Wolf came to Gluskabe. Wolf looked very much as wolves do to this day. 'Alnobak,' Gluskabe whispered.

"'If they walk their way,' Wolf said, 'I will walk mine.'

"'That is good,' Gluskabe said.

"Other animals came up, one by one. When Gluskabe whispered the name of the human beings to them, to the deer, the caribou and the elk, they all gave the same answer. 'We will stay away from them. If they come to hunt us, we will run away.'

"'That is good,' Gluskabe said.

"When Gluskabe whispered the name 'Alnobak' to Mahtegwas, the Rabbit, Little Mahtegwas was so frightened that he began to run around in circles. But not all of the little animals were frightened. Some of them, like Chipmunk and Deer Mouse, told Gluskabe that they would live as close to the human beings as they could so that they could eat the food which the human beings dropped on the ground.

"'That is good, too,' Gluskabe said.

"At last, Gluskabe thought he had spoken to all of the animals which had gathered. A few animals, some of the ancient fierce ones, had not come when Gluskabe called. That made Gluskabe sad, for he knew that it meant they would hunt the human beings. Some day, all of those fierce ancient ones would have to be destroyed.

"As Gluskabe thought about those dangerous monsters, he looked up and saw one more animal sitting, patiently, at the edge of the clearing. This animal looked a bit like Wolf, but was different. As Gluskabe looked at that animal, it came trotting up and then sat at Gluskabe's feet. It was Dog.

"'Master,' Dog said, 'I know about the ones who are coming. I have been waiting for them.'

"'What will you do when they arrive?' Glusklabe said. 'Do you intend to hunt them or bother them in any way?'

"Dog laughed. 'Master,' Dog said, 'I want to live with them. I want to sleep by their fires and share their food. I want to take care of their children. I will help them when they go hunting. If there is danger, I will warn them and I will risk my own life to save them. I will be their greatest friend.'

"Gluskabe looked into Dog's eyes and Dog looked back at him. Gluskabe could see that every word which Dog spoke was true.

"'My friend,' Gluskabe said, 'it will be as you wish. You will go and live with the human beings. Even though some of them will not deserve to have such a great friend, you will be loyal to them and sleep by their fire. That is how it will be.'

Sings Like a Thrush rubbed her hands together and then spread them out over the fire.

"That is what happened back then," she said. "That is how it is to this day."

She looked down to her side where Hummingbird smiled up at her as she sat with her puppy. Although she was struggling to stay awake, the eyes of her daughter were filling with sleep. But the eyes of Awasosis were bright and awake. The little dog looked up at her with an alertness that, for just a moment, surprised Sings Like a Thrush.

"Ah," Sings Like a Thrush said, "so you have heard the story and you understand it, too, Awasosis?"

The puppy licked her hand as she held it out. Then she curled up next to Hummingbird, whose eyes were closed now, but who still reached out one arm to draw her dog in close to her.

Sings Like a Thrush pulled the deerskin blanket over the two little ones.

THE TRACKS OF THE GIANT BEAR

MUSKRAT AND KWANIWIBID FOLLOW THEIR NOSES

"Moskwaso, kina! Muskrat, look here! I was attacked by some one with no arms and no legs who was a great maker of songs. He had no club, but he still tried to hit me. Now I have defeated him. Do you know who he is?"

Muskrat looked up from combing the burrs out of the fur of the big dog that lay contentedly on its back, front legs straight up in the air. Although he was bigger than any other dog in the village and it was three full winters since he had left his mother, Awasosqua, in the village of the Salmon People, Kwaniwibid, or Long Tooth would still roll over like a little puppy when he felt the strong, slender fingers of Muskrat stroking his chest. He would open his mouth and let his tongue loll over those two long canine teeth of his which had given his name. His teeth were longer than those of any other puppy in the village. Teeth like that were thought to be a sign that a dog would be a good hunter.

A very old man stood in front of them with his hands behind his back. It was Tomakwa, one of the old people in their village who was great at making riddles. He had reached the age, Muskrat understood, when making riddles was all that he wanted to do. Tomakwa would wander around all day thinking of ways he could say things that would make sense but which would be hard for the people to understand. Even though the Day Traveler was only a hand's height up into the sky, Old Tomakwa had already found some kind of riddle to bring.

Muskrat could tell from the sweat on the old man's forehead that whatever he had behind his back was heavy, and it was not easy to hold.

He tried to look, but Tomakwa stepped back and turned so that the boy could not see.

"Kwaniwibid," Muskrat said to the dog, "what is our grandfather holding there?"

Kwaniwibid looked up and sniffed the air. Then the big dog looked straight at Tomakwa as the hackles on his neck rose and he made a soft deep growl, the warning sound that danger was close by.

"I have the one I am talking about right here," Tomakwa said. "Can you not guess who he is?"

As soon as the old man said that, Muskrat knew. But he also knew how much delight old Tomakwa got out of fooling others, so much delight that he would even risk his life to do it, it seemed! Muskrat smiled. If he could be like Tomakwa when he was very old, then growing old would not be such a bad thing at all. Only the children played as much as old Tomakwa did.

"My elder," Muskrat said, "I cannot guess. You must show me that one you are talking about."

Tomakwa pulled one arm from behind his back and quickly swung his other arm—which had been holding the rattles to keep them quiet—around to help him keep his hold on the neck of a huge rattlesnake. The snake was as big around as a man's arm and it was not happy. Its mouth gaped open to expose fangs that seemed as long as a child's finger. Even more sweat appeared on Tomakwa's brow, but there was a happy grin on his face.

"I have fooled you, young man!" Tomakwa said.

"That is true, my elder," said Muskrat, his hand on the neck of Kwaniwibid, whose growling was even deeper now. "But it seems to me that the one you are holding is ready to return to his home up in the stones on the bare hill. Would you like me to go with you to help bring him home safely?"

"Young man," Tomakwa said, "do not trouble yourself." As he spoke, the snake wrapped its length around his left arm and now the old man's two arms were held together as firmly as if they had been bound by sinew cord. "However," Tomakwa said, "if you have nothing else to do, I would enjoy your company."

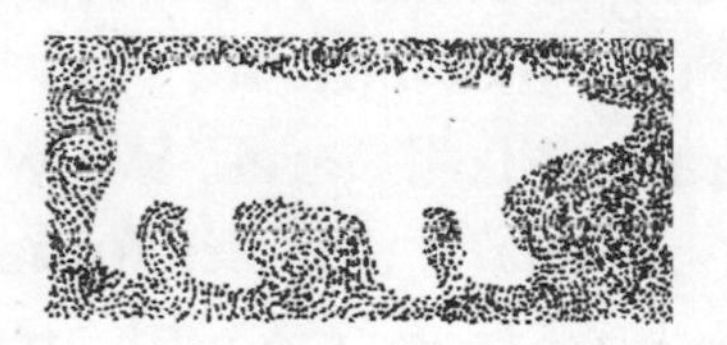

When Muskrat returned, his big dog was still sitting, guarding the small lean-to of hemlock branches where Muskrat had taken to spending his nights. Now that the tall, slender boy had reached the age of a dozen winters he no longer had to sleep every night beside the fire of his parents. So he had made this little summer shelter for himself and even managed to feed himself now and then from the game and fish he would catch—with

Kwaniwibid's help. However, his lean-to was less than a look away from his parent's lodge, close enough to smell his mother's cooking and know when it was time to pay them a visit and make sure there was no food left over.

"Kwaniwibid," Muskrat said, "you have done well. The brave one rattled his tail once and then crawled down into his lodge in the rocks. Tonight he will tell his children how he defeated an army of strange giants with arms and legs. And Tomakwa is off looking for another riddle to bring to someone. But I think this time it will not be a riddle that might bite him."

Muskrat flopped down and put his head back onto Kwaniwibid's shoulder.

"My friend," Muskrat said, "I think it is time for us to take the winter trail and try to find those strange tracks again."

Kwaniwibid put his paw on the boy's arm and Muskrat turned to look into his dog's eyes.

"Do you not agree?" Muskrat said. "Do you think there is danger there?"

Kwaniwibid wagged his tail and it thumped hard against the earth, making a sound like a drum.

"Good," Muskrat said. "I agree. We will be careful, but we are not afraid. Remember that story old Tomakwa told us last winter about the hunter whose dogs saved him when he was chased by the great beast? When a human being and a dog take care of each other, nothing can defeat them!"

Muskrat jumped to his feet in one quick movement, grabbing up his spear thrower in one hand and four spears in the other. His spear thrower was an arm's length grooved piece of wood carved so that a spear would rest in it. Scattering the hemlock boughs which were laid down on the floor of their lean-to, Kwaniwibid, too, scrambled to his feet. Within the space of a few heartbeats, the two of them had disappeared up the trail which led into the hills above their village and toward the Forever Winter Land.

There was no boy and his dog who could run as fast or as long as the two of them. By the time the Day Traveler was in the middle of the sky, they had gone farther than most grown men would travel in a full day. They had reached the mouth of the high valley, the place where they could feel the cold breath from the ice river caught up there between the hills.

In the days of Muskrat's great-great-grandparents, the ice river had been longer. Each summer it grew smaller and held less of the winter in its shaded valley. In the time of the ancient ones, such rivers had been everywhere on the land and the People of the Dawn had been forced to move toward the summer lands. But that was long ago, before the hills had grown tall. Still, few people came to this valley. The hunting was not good here and only a boy searching for adventure would find a reason to come to this place and find, as Muskrat had done three dawns before, tracks like those of a bear, yet unlike any other bear tracks he had seen before.

It did not take them long to again find the tracks in the soft gravel where the melting ice flowed down making a little stream. But the old tracks had been joined by new ones, ones so fresh that small grains of sand still were crumbling from the edges. Muskrat crouched to put his hand down and spread it over the clearest print. The track was wider than the span from the end of his thumb to the end of his littlest finger. The track led back up onto the ice. Muskrat listened. All he could hear was his dog's breathing and the soft, friendly trickle of the little stream.

Muskrat narrowed his eyes and shaded them with a hand.

"Kwaniwibid," he said, "I do not see anything there, but it must be close. It is all white in the valley and surely the dark coat of a bear would be easy for us to see."

Kwaniwibid growled. It was a deeper, throatier growl than he had made when Tomakwa held out the rattlesnake.

"Remember," Muskrat said, "in the old story, it was the dog who smelled the monster first. So be watchful, open your nose as we go forward."

As soon as he took his first step onto the ice, Muskrat could no longer hear the comforting sound of the trickling stream. The sound of his own heartbeat had grown so loud that it seemed as if he was hearing distant thunder. He took a few deep breaths and the sound of his heart graduallly grew softer.

Kwaniwibid looked up at him, his head cocked to the side and a concerned look in his eyes. Muskrat said nothing, but motioned forward with the hand that held the four spears. Kwaniwibid lowered his head and began to walk, ears pricked forward, tail straight back.

As they climbed through the valley, which narrowed as it went higher, Muskrat noticed the way the dark heads of the stones lifted out from the ice like sleepers sticking their heads up from under a white blanket. The surface of the ice near the sloping face of the hill was rough and not slippery to walk on, even though it seemed as smooth and shiny as the surface of a pond in the center of this little valley, no wider than two spear casts. The ice felt as old beneath his feet as the rocks themselves. The wind was blowing from behind them now, seeming to push them forward.

Our scent is going ahead of us, Muskrat thought. It worried him, but there was no other path to take unless he turned back. He had seen one more track in a place where a slant of sun through the two peaks had softened the snow, and he knew the bear was somewhere ahead of them. But none of the boulders were large enough for it to hide behind. And he would see the dark fur of a bear in this white valley, even though it was so bright here from the reflected sun that he still had to squint his eyes to see things clearly.

Suddenly Kwaniwibid stepped in front of him and stopped. The big dog lifted his head, trying to sniff the air, the hackles on his back rising. That other sense, the one beyond sight and smell and hearing, was speaking to the dog. In his own way, Muskrat felt it, too. They were being watched.

Muskrat looked carefully ahead of them. The slope was smooth except for one hill of snow just ahead of them. Suddenly Muskrat saw that the surface of that hill of snow was being rippled by the wind, trembled as fur trembles.

The hill stood up in front of him as the tall boy saw it was no hill at all, but a great bear that was white as the snow.

Kwaniwibid yelped twice and dove at the bear's flank, drawing it toward him. Muskrat tried to fit a spear to the spear thrower as he drew his arm back, but his hand was trembling.

The bear was making a rumbling sound now, like the sound of boulder rolling down hill, as it circled and struck at the dog that was too quick for the bear to catch.

Muskrat stepped closer and the bear turned its long head toward him. Its head was flatter than that of a dark-furred bear, almost as flat as the head of a rattlesnake. It raised a paw and struck.

But before the bear struck Muskrat had leaped away, leaped toward the bear's right side as the voice of old Tomakwa's story reminded him that—"a great bear always strikes with its left paw."

As he leaped, Muskrat lost hold of his spears and his atlatl and they went flying down the slope, skimming like swallows over the slippery face of the ice slide at the center of the valley.

The bear reared up onto its hind legs and Muskrat saw in that moment how thin it was, how its fur was almost yellow on its belly and how one of the teeth it bared as it roared was broken. He saw also what he had to do. Grabbing hold of Kwaniwibid with both arms, the tall boy somehow lifted the big dog off its feet. Then, before the bear could come back down on all fours on top of them, Muskrat had taken the few steps that took him to the ice slide and stepped onto it. His feet shot out from under him and he fell onto his back so hard that his breath exploded from him and he saw the lights of small stars. Yet he held on to Kwaniwibid and did not let go as the two of them went hurtling down the long slippery slope.

Somehow, Muskrat managed to look back as they slid and he saw that the great white bear was not following them. Instead, it had turned away, as if losing interest and was moving slowly, step by step, up higher toward the top of the ice hills. Then Muskrat's head struck something hard and he saw nothing else.

When he woke, Muskrat felt moisture on his face. It was warm and for a moment he thought it was blood. Then, for another heart-stopping moment, he thought that the breath he felt on his cheek was that of the great white bear and not that of his big dog as Kwaniwibid leaned down to lick the boy's face again.

Muskrat sat up and felt himself all over. There were plenty of sore places, but all his bones seemed sound and his arms and legs worked well, even if every breath he took hurt his sides. His atlatl and spears were close by in the gravel at the base of the ice. He washed his face and hands in the chilly waters of the little stream before picking up his weapons. He knew he would not need them.

His big dog was looking back up the slope. As Muskrat came to him, Kwaniwibid whined and turned his head toward the hill of ice.

"Nda," Muskrat said, "that old one has come here to die. It is walking back toward the winter land that gave it birth." Muskrat smiled. "It is like what happened to old Tomakwa with that rattlesnake. It would not have been a problem for him if he had not picked it up. That great bear would not have troubled us if we had not followed it."

He patted the big dog's head. "But it has given us a story to remember it by."

Then Muskrat and Kwaniwibid turned back toward the summer lands. Though both of them limped as they went, the two of them walked back toward their own home more wisely than they had walked before.

DOG PEOPLE

KEEPS-FOLLOWING-THE-TRAIL MEETS SOKSEMO'S PEOPLE

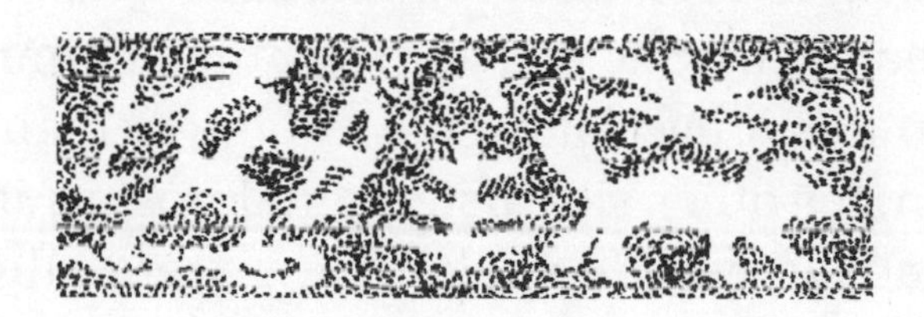

So that is how Snowy Owl defeated the great beasts that were so big they were like walking hills. Then he and the last daughter of Great White Hare lived together happily in that village in the North Land. They were still living there happily when I left them."

Stands-in-a-Hole waved one hand over the fire and then leaned back and took a deep breath. The six children gathered around the fire in front of the short old man remained as silent as a family of mice who think they have heard the call of a hawk.

Now, Stands-in-a-Hole thought, *my great grandson Muskrat, the son of my daughter's daughter, will rub his nose and shake his head as if he has just woken up.*

Sure enough, just as the old man expected, at that very moment Muskrat stretched out one arm and, much like the Muskrat whose name he shared, he rubbed his face and then blinked his eyes.

Now, Stands-in-a-Hole said to himself, *Hummingbird will smile.*

Indeed, just as his mind's voice told him it would happen, Hummingbird opened her mouth to smile and then laugh. The voice of the little girl from the Salmon People Village was almost as high as that of the tiny swift-winged flyer. Then she hugged her three-winters-old dog, Awasosis. Although the dogs that were the companions of the other children were faithful to them and seldom far away, it seemed as if little Awasosis was always at the side of her dearest human friend.

Awasosis was even closer to Hummingbird than the girl's best friend, Cedar Girl. She had accompanied Hummingbird and her parents on this long visit to the village of their cousins. Cedar Girl sat on her other side listening almost as intently to every word Stands-in-a-Hole spoke. They were all the guests of Walimogwkilskwasis, Sweetgrass Girl, whose parents owned this lodge and were sitting close together, their backs against the southern wall. Little Sweetgrass Girl listened closely to Stands-in-a-Hole. But the old man noticed, with great amusement, that she also was always turning her ears in every other direction as well, trying to catch every word anyone spoke—especially if that word was meant to be secret. Stands-in-a-Hole looked at Rabbit Stick, noticing how the boy's eyes kept moving toward Sweetgrass Girl. Sweetgrass Girl made it very clear that she did not notice Rabbit Stick. She made it so clear that the old man laughed inside. *Those two,* he thought, *will surely marry some day.*

And, Stands-in-a-Hole continued, looking at the boy sitting next to Rabbit Stick as he talked to himself inside his head, *of course, Naboomsawinno, Keeps-Following-the-Trail, will ask if it is still that way to this day.*

"Great Uncle who is the brother of my grandfather," Keeps-Following-the-Trail said, raising up his left hand to touch the tips of his fingers to his lower lip, "are any of those great walking hills still left alive? Does the Great White Hare still live in the North Land? Are there still such great storms there which can turn a person into ice?"

Stands-in-a-Hole tried not to laugh. He had learned long ago that it was not a wise thing to let others know how often he could tell exactly what they were thinking. Still, it gave him pleasure to be able to do that. It had nothing to do with his powers as a deep-seeing man, one who could sometimes foretell the future. It was just that he paid close attention to all the people around him in his village and in the other villages of the Dawn Land People. And he had a very good memory. That was why he was also known as Great Storyteller—a name he somewhat preferred to Stands-in-a-Hole. And his good memory told him that he would surely have to answer the questions asked him by Keeps-Following-the-Trail, whose name was a good description of his determined nature. Still, he would tease him a bit first.

"Ah," Stands-in-a-Hole said. "How would I know such things?"

"You said in your story that you were there, Great Uncle," Keeps-Following-the-Trail said. "Surely you must have seen those things."

"Hmm, that may be so, but didn't I just hear the rumble of thunder outside?"

"Nda!" All six of the children spoke as one. The thunder had not sounded yet from the sky. And the sound of spring thunder—which was a voice from the sky telling the people to stop telling such old stories until the cold time came again—had certainly not been heard that night.

Again, Stands-in-a-Hole almost laughed, but he kept his face straight.

"Well," he said, "if I did not just hear the thunder, then I suppose I can continue to tell these stories and talk about them with you. So, second son of my brother's son's son, I will answer you. It is said in the story that Snowy Owl and Little One destroyed all of the monsters as big as walking hills, but perhaps they only destroyed those which were in one herd. As you know there are many herds of deer and more than one herd of caribou. So, there may be more of those great beasts in the North Land. And as to Great White Hare, we see his little cousins in our own land, so it would seem to me that he must still be there, even though his powers are not as great as they were back then."

Stands-in-a-Hole nodded and again spread his hands out over the fire. He knew that he had left one question still unanswered, but he wanted to see if—as he expected—Keeps-Following-the-Trail would continue to follow the trail.

"Ktsi Nudatlogit," Keeps-Following-the-Trail said carefully, "Great Storyteller, what of the big storms that could turn people to ice?"

Stands-in-a-Hole could not stop himself from smiling this time. "Great Nephew," he said, "I am glad that I am not the deer which has you on its trail! Yes, those storms can still come. True, the winters are not like they were when I was your age. Still, it is wise to always be prepared for the wind's breath to change."

Keeps-Following-the-Trail nodded and pressed his fingertips again against his lips, as if eating the old man's words to make them a part of him forever.

Stands-in-a-Hole rose to his feet. "You have all listened well," he said. "If the thunder does not speak, perhaps when the Day Traveler sleeps tomorrow night I may again tell you something of how it was long ago."

"Onnh-honnh," the six children said as one. "Oleohneh, Ktsi Nudatlogit."

Then, without having to duck his head, Stands-in-a-Hole walked out of the door of the wigwam toward his own lodge.

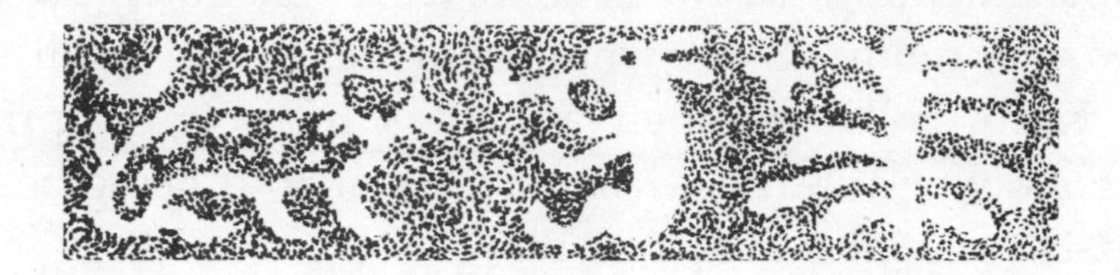

The next day, though it was earlier in the season than any had expected, the thunder spoke. A warm wind came from the southland and swirled toward the sunrise and the great waters. Then a rain swept over the village that was so hard it washed away every bit of the remaining snow in the woodlands as thoroughly as a spring-swelled stream carried away pieces of bark caught in its flow. The voices of the thunder beings rolled and growled overhead, and the children did not gather that night in the lodge of Sweetgrass Girl to hear stories. But they were not disappointed. The sweet smell of the dark earth and the roots of the new grasses and this new season of promise were stories enough for them.

Two sunrises later, Keeps-Following-the-Trail walked toward the north with his dog, Soksemo, ranging ahead of him. Soksemo was almost as large as his brother, Kwaniwibid, and was even better at following the scent of an animal, even if that scent was very old. Soksemo, Good Nose, looked much like his dark-furred brother except for the circle of white fur like a necklace Soksemo had around his neck. So, even at a distance, it was easy to tell the two dogs apart.

Keeps-Following-the-Trail wished that his friend, Rabbit Stick, had agreed to come with him. It was rabbits he was going to hunt, and of all the boys Rabbit Stick had the best throwing arm. But Rabbit Stick preferred to stay in the village to tease Sweetgrass

Girl. He would follow her about, making jokes and tossing pieces of bark at her until she would finally turn and—in many cases—hit him with something heavy. For some reason, Rabbit Stick thought this was very funny, and he would show his bruises or his bloody nose to Keeps-Following-the-Trail with something like pride. But Keeps-Following-the-Trail was not interested in being beaten up by girls. He was interested in hunting.

Ahead of him, Soksemo whimpered. Keeps-Following-the-Trail looked at the earth made soft and wet by melting snow. It was the track he had hoped for—not that of a small rabbit but the bigger print of a hare.

"Go," Keeps-Following-the-Trail said, "bring him around to me."

Soksemo looked up at Keeps-Following-the-Trail with the familiar look in his eyes which always showed his pleasure. He pressed his nose against the boy's hand and then trotted up the trail that disappeared among the rocks. Keeps-Following-the-Trail leaned back against a tree. The clear space in front of him was where the hare would run, making a big circle as it tried to outdistance the dog. His rabbit stick would whirl out and strike it—if he was lucky. Keeps-Following-the-Trail took one slow breath after another, watching. Then he saw it. The white hare was coming through the trees. It was easy to see for its coat was still white to blend with the winter snow; it had not yet turned color to match this new season. It was bigger than any hare he had seen before! Keeps-Following-the-Trail clutched his stick tighter. As he did so, the white hare turned and ran straight back through the trees. It was not coming any closer now and it was too far away for a throw.

Keeps-Following-the-Trail whistled for Soksemo.

"My friend," he called, "he is running from us!"

Then Keeps-Following-the-Trail began to run after the hare. Soon Soksemo caught up to him and then led the way. They ran for a long time, heading straight toward the north. As they ran, the boy noticed the air was suddenly growing colder. The wind had shifted and was now coming from the direction of the winter land. The wind grew in strength as they ran. Suddenly its touch was so firece and cold that it seemed as if the fingers of Old Winter Man himself were slapped across his face. For a heartbeat, Keeps-Following-the-Trail thought of stopping and turning back. Then he saw the white hare, standing on its hind legs and looking back at them from the hilltop just ahead.

Soksemo looked up at Keeps-Following-the-Trail. There were small flecks of snow in the air now, like the down feathers of a goose. The boy looked down at his dog and saw that his friend was ready to stop the chase if that was what he wanted.

"Nda," Keeps-Following-the-Trail said. "We will catch this one!"

The white hare whirled and disappeared over the hill, and the boy and the dog followed.

So it went on for a long time. The snow and the wind grew heavier, and the hare continued to run just ahead of the two who pursued him. Then, as quickly as a single heartbeat, it vanished.

Keeps-Following-the-Trail stopped. He could no longer see either the hare or the trail it had been leaving in the snow. The snow was so heavy and the wind so strong that everything around them was white. He could not even see Soksemo. Suddenly he realized how cold he was.

"My friend," Keeps-Following-the-Trail called, wondering if he was both lost and alone. But as soon as he spoke the familiar shape of his dog appeared by his knees and he felt Soksemo's warmth press against his legs.

Keeps-Following-the-Trail knelt. "My friend," he said, "Soksemo, I have been foolish. The white hare has tricked us. I am afraid that we will freeze here."

The dog pressed hard against him with its body, making him take one step backward and then another. Soon they were in a place where the snow had piled itself deeply, making a great drift between two huge stones where a hemlock's branches bent low. Then Soksemo began to dig, disappearing into the snow and then turning to look back at the boy.

"You are right, my friend," Keeps-Following-the-Trail said, "we must make a snow cave."

He joined the dog in digging, breaking off hemlock branches to hold the sides of the cave in place. Soon it was just large enough for the two of them to curl up on the mat the boy made with more hemlock branches. The mouth of their cave faced away from the wind, and when Keeps-Following-the-Trail had placed the last of the wide-needled branches as a door, it seemed almost as warm as it would be in a lodge. Yet Keeps-Following-the-Trail felt the numbness in his hands and feet. They were like ice. He wished he had worn at least a deerskin shirt and not come on his hunt bare-chested. Soksemo pressed against him to share his warmth. The boy buried his fingers in the big dog's thick fur and felt the feeling return to them. He listened for a while to the muffled sounds of the storm outside their cave. Then, because there was nothing else to do, he fell asleep.

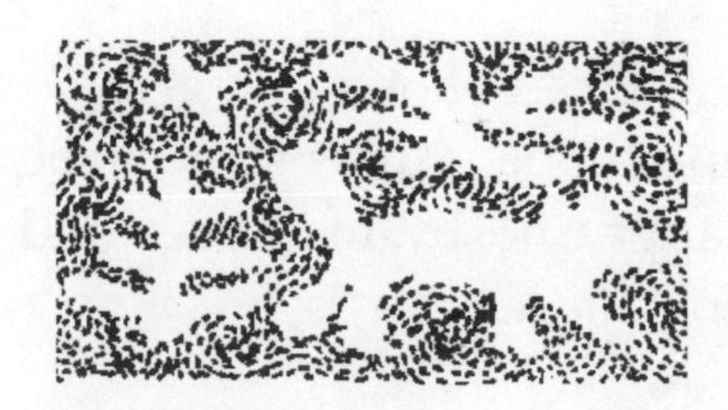

When he woke, the storm was gone and so was the snow cave. Soksemo was nowhere to be seen. Keeps-Following-the-Trail was in a small lodge he had never seen before. A young man was bending over him. The young man's face looked friendly and he wore a white necklace made of animal bones strung together. His eyes were a strange color of brown, but they were filled with concern.

"My friend," the young man said, "I am glad you are well. I am your friend. My people told me to bring you to this lodge where you would be safe. When we are in your village, we appear to be your dogs. But now that you have come to visit us, you see us as people. You see us just as we see each other and as we see ourselves. Now that you are awake, come out and meet the others."

When Keeps-Following-the-Trail came from the little lodge he found himself in the middle of a village. He had never visited this village before, but everything that he saw around him seemed to welcome him. The Dog People, who looked like humans dressed in different colored animal skins, were all around. All of them pressed close to touch him and greet him. Though he had never seen them before in their human shapes, he knew that they were his friends and truly wished to help him.

"Come," said the young man with the animal bone necklace, "my people have a meal for you. Sit by the fire and enjoy yourself."

Keeps-Following-the-Trail did as his friend told him to do. Even though his friend's voice was almost like a growl, he liked the sound of it. He knew he had heard that voice before. He sat and ate and then listened as the people entertained him with songs and stories about hunting. All of the people in this village loved to hunt, the women as well as the men. It felt good to be with them. He watched them closely, trying to figure out what they looked like when they were again in the form of dogs. Perhaps that fat young man there wearing a brown skin was actually Moosis, the special friend of Sweetgrass Girl. Perhaps that very small person with the gray-colored skin was little Mikwe, the companion of Rabbit Stick. But who was the young man who had welcomed him? Keeps-Following-the Trail smiled. He was certain that he knew.

"Remember," the Dog People told him, "only hunt when you need food for your people. Always say thanks to the animals that you hunt and you will always have success. That is the way we hunt."

"I will remember," Keeps-Following-the-Trail said.

At last, he grew tired. His friend saw this and took the boy by the arm.

"Come," his friend said, "back into my lodge to sleep. When you wake, you can go home."

Again, Keeps-Following-the-Trail closed his eyes and slept. But when he woke, he was no longer in the lodge of his new friend in the village of the Dog People. He woke because someone had reached through the wall of his snow cave to grasp his arm and shake him. Beside him, Soksemo was whimpering and digging at the wall of their shelter.

Stiffly, Keeps-Following-the-Trail sat up and saw that the arm belonged to old Stands-in-a-Hole who pushed the rest of the way into the snow cave which was falling apart around them. Behind the old man was Keeps-Following-the-Trail's father and other men of the Only People.

"The boy is all right," the old man called, in a bigger voice than it seemed his short body could hold. "Even though he has been missing for two days, it was just as my seeing told me. He saved himself from the great storm in this cave."

Though he felt weak, Keeps-Following-the-Trail stood up, leaning down onto the old man's strong shoulders. He looked down for a moment and saw Soksemo looking up at him. There was a different sort of knowing look in his large brown eyes. Keeps-Following-the-Trail smiled as he saw the familiar circle of white fur around the big dog's neck. It looked much like a necklace made of animal bones.

"Nda," Keeps-Following-the-Trail said, "I did not save myself. My friend saved me."

LOST IN THE SNOW

RABBIT STICK AND MIKWE HELP SWEETGRASS GIRL AND MOOSIS

Like the small footsteps of a bird, the days were now growing longer. Still, there was much of the winter time ahead. Cedar Tree, the father of Sweetgrass Girl, was glad of it. More than anyone else in the village of the Only People he liked this time for he was able to walk far on top of his snowshoes and see the land in new ways. He was happy whenever he went out to check on his trapline, often with his wife Singing Beaver walking beside him to help pull their toboggan.

Sweetgrass Girl did not always go with them. Sometimes, as she was doing this morning, she would curl up in her warm bearskin robe and stay in the wigwam of her special friend, Hummingbird. It was true that she was only the third best friend of Hummingbird. First came Awasosis, Little Bear, the dog who was always by the side of little Hummingbird. Second came Cedar Girl, who often came to visit from the village of the Salmon People, especially in the season when the earth was covered with the Old Winter Man's cold blanket. Third came Sweetgrass Girl.

But Sweetgrass Girl did not mind Hummingbird's friendship with Cedar Girl. Cedar Girl was not always in the village, and even when she was there she was such a bright and likable person that Sweetgrass Girl thought of her as one of her own best friends, too. Everyone liked to have Cedar Girl around. As far as Awasosis was concerned—Sweetgrass Girl had her own dog, one of the brothers from that same special litter of six. And like the other five children who had adopted those puppies as their relatives and friends, Sweetgrass Girl knew that she would never have a friend more faithful than Moosis, Little Moose. Even now, his head back, all four feet up in the air above his very fat, wide belly, Moosis lay by her side, right next to Awasosis whose head rested on the lap of Hummingbird. And on the other side of Hummingbird Cedar Girl was snoring, her arms around her skinny dog Azeban.

Of all the human people and dog people in the big wigwam of the family of Hummingbird, only Sweetgrass Girl was awake. It was often that way. She did not seem to need as much sleep as others and she was glad of it. It meant she could listen to that much more and hear that many more things that she could remember and talk about later. And when there was nothing to hear, she could think about the things she had heard and just how she would tell someone else about it.

Right now she was imagining how she would tell Hummingbird and Cedar Girl—if her lazy friends ever woke up!—about the argument old Pitch Hand had with his wife Turtle Woman the night before. Sweetgrass Girl and Moosis had been so shocked when they heard them fighting, that all they could do was just sit there quietly under the cedar tree near their lodge. It seemed that once again old Pitch Hand had borrowed something from someone else in the village without telling the person and his wife was insisting that he take it back. But in the midst of their quarreling, whatever it was that he had taken was broken. It would be a funny story to tell.

Telling stories and hearing them—that was what Sweetgrass Girl loved most. It was another reason why she stayed in the village instead of going out with her parents. It was nearing the end of the time of long nights and short days. In only another moon, the storytelling season would end. Then the stories—which old Stands-in-a-Hole insisted were not just words spoken, but people themselves—would travel back to their own lodges in the Always Winter Land. There the stories would sit around their fires telling themselves to each other. If Sweetgrass Girl stayed in the village, she would be able to sit with the other children and their dogs around the fire and hear Stands-in-a-Hole tell more of those ancient stories. Her mother, Singing Beaver, was a good storyteller, and Cedar Man could relate many tales about hunting and trapping, but no one among the Only People could tell a story as well as Stands-in-a-Hole. So, whenever the old man told stories, she tried to be there, listening and remembering. Secretly Sweetgrass Girl promised herself that one day she would also be called Great Storyteller by the children of another generation.

When no one else was around, Sweetgrass Girl practiced her storytelling on Moosis. It was not an easy thing to do. Though he was not as tall in the shoulders as his bigger brothers, Moosis was the heaviest of all the dogs in the village, as fat-bellied as a moose that has been filling itself with tender water plants for a whole summer. Whenever Sweetgrass Girl told him a story in her soothing voice, Moosis would try hard to listen, but sooner or later his wide head would drop and he would start to sleep. “Kita!” Sweetgrass Girl would shout. “Listen!” And Moosis would jerk his head up with a guilty look in his eyes.

That night, just as she had hoped, Stands-in-a-Hole came to tell stories. The circle formed about him was the usual one, those six children and their dogs: Hummingbird and Awasosis, Cedar Girl and Azeban, Keeps-Following-the-Trail and Soksemo, Muskrat and Kwaniwibid, Sweetgrass Girl and Moosis, and that pesty boy Rabbit Stick and his little dog Mikwe, which had been the smallest of the litter and the last to be adopted. Even though Mikwe was four winters old, he was half the size of his biggest brother. He seemed to be the brightest of all the dogs, but Sweetgrass Girl still wondered why Rabbit Stick, Mahtegwasabazi, who was growing quite tall and strong now, should have chosen to care for such a small one. For two winters now he had been pestering her. True, she never actually told him to leave her alone, but one would think that being hit with a stick when he tried to tickle her would be clear enough!

Sweetgrass Girl suddenly noticed that Rabbit Stick was looking straight at her as she looked at him. She felt her face grow very warm—perhaps she was leaning too close to the fire—and she turned her eyes toward another part of the big lodge, trying to make it seem that she had just been looking around, not staring at any one person.

She looked up at the poles that went from one side of the big lodge to the other. Unlike many of the other lodges in the village, this wigwam was not shaped like a rolled cone of bark but was flat-walled and long, like several wigwams joined together. It was not an easy lodge to move, like the other smaller ones, but it could hold more people. And that was good, for in addition to Hummingbird, her family and their dogs, including Awasosqua, the mother of the six young ones, all four of her grandparents and the families of her mother's two sisters lived here. People joked that visiting their lodge was the same as making a visit to another village. Sweetgrass Girl found herself wondering what sort of lodge she would have when the day came for her to marry, when with her husband by her side she would look down on the beautiful face of her own grandchild. Once again she felt her face grow warm, even though she was now leaning back from the fire.

"Would you hear a story?" It was the voice of Stands-in-a-Hole which spoke. It brought Sweetgrass Girl back from wherever she had been, but for once she was not the first of the six friends to answer the old man's question with a loud yes!

"Unnh-Hunnh," said the five other children, with Sweetgrass Girl's voice close behind.

Stands-in-a-Hole looked straight at Sweetgrass Girl for what seemed to her a very long time, even though it was only the duration of a few heartbeats. Then, a small smile at the edge of his mouth, he began to tell the tale of the adopted boy who was raised by his grandparents.

Four sunrises passed and the parents of Sweetgrass Girl did not return. One storm after another had rolled in, piling snow up so that the people had to use thick pieces of elm bark to shovel out the entrance to their lodges each morning. Two wigwams in the village had to be moved to keep from being drifted over. Trackers went out from the village, but so much snow had fallen that they could not find their trail. They found the marker tree where Cedar Tree had scraped with charcoal the rough figures that indicated he and his wife had gone upstream to check his beaver snares and would return after two sunrises, but beyond that there was no sign. Two more sunrises came and went and now there was real concern in the village. That night, Sweetgrass Girl went to the lodge of Stands-in-a-Hole.

"Grandfather of my uncle," she said, handing him her gift of dried muskrat roots, a medicine she knew the old man could make use of, "I want to know if you have seen anything of my parents."

Stands-in-a-Hole sat her down beside his fire and spread out a deerskin that had been marked with all kinds of shapes. There were circles and twisting lines and birds opening their wings. Stands-in-a-Hole took the pouch from his side and shook out a handful of white carved bones onto the deerskin. As he breathed in and out, in and out, he looked hard at the way the bones had fallen. Sweetgrass Girl held her breath.

"They are still living, I think. But they are in trouble. What the bones tell me—and what my heart knows of your parents—is that they did not continue farther up their trapping line when the storms began to come. Instead, they would cut this way, five looks away from here, angling down toward our village. They would make for the High Ridge caves, three looks away, to find shelter. That is what my heart thinks and what the bones also tell me."

All through that night, Sweetgrass Girl made things ready. Near dawn, she bundled food and two fur cloaks into packs. One pack was tied to the back of Moosis, who sat patiently as if knowing exactly what they were going to try to do. She readied her stockings made of the skin of the white hare and put on her high boots, each made of the skin taken from the hind leg of a moose. She pulled on her mittens made of beaverskin. Then, just as the first light of dawn was about to appear, she reached for the skin which covered the door of the lodge. But before she could touch it, someone scratched on the wall of the wigwam by the door.

Excitedly, thinking it was her parents, she flung open the door. The light from her fire reflected on the tall, rangy boy standing there. He was dressed in heavy clothing made of beaverskins that hung down past his knees to feet also covered with mooseskin boots. His little dog, Mikwe, stood behind him, a white cloud of breath about his face.

"I will go with you," Rabbit Stick said. For once, there was no joking in his voice as he spoke to her. Somehow, he looked older than the twelve winters both of them shared.

"Do not get in our way," she said in a voice that sounded much angrier than she felt. In fact, though her heart still feared for the safety of her parents, she felt something like a bird beginning to flutter its wings in her chest.

"I will be behind you," Rabbit Stick said. The calmness of his voice made her stand straighter. The uncertainty she had been feeling, a weight that bent her shoulders, began to leave her. She pushed past him. Half a step behind, Rabbit Stick followed her toward the trail that led in the direction of the cliffs.

When the snow grew deeper, it was Moosis who took the lead. His wide, heavy body broke through the snow, making them a passage, and it seemed as if nothing could tire him. No snow was falling now and they could see everything around them with great clarity as the Day Traveler began his journey across the Sky Land. Little black-capped singers were calling around them and the promise of warmer seasons was in the air. No one else in the village had seen them go or the older men would certainly have followed, even though all of the best trackers in the village were exhausted from the days of searching just past.

Now the Day Traveler was four hands high, and they had covered much distance, more than two looks. As they started up the steep slope, where rocks were exposed by the wind, Sweetgrass

Girl's foot slipped. She stumbled and almost fell. But, just as quickly as she lost her balance, she found herself pulled back. Moosis had grabbed her heavy cloak, which she had belted tight around her with a sinew cord, with his teeth. And the strong hands of Rabbit Stick had her by her arm at almost the same moment.

"I am fine," she said, shaking his hands off and then petting Moosis. "Good dog," she said. Then she started forward again. After a few steps, she began to wonder why she felt lighter. She looked back and saw that, as he pulled his hands back, Rabbit Stick had also pulled the heavy pack from her back. It was over his shoulders now, made lighter by a tumpline he had fastened to go over his forehead.

She started to say something and then stopped. It would be too much trouble to make someone so stubborn act sensibly. She satisfied herself with a loud "Hummph!" Then she turned back to following Moosis up the trail, not noticing the smile on Rabbit Stick's face.

When they reached the place where the trail turned toward the caves, Sweetgrass Girl stopped and her heart sank. She understood now why her parents had not returned if they had made their way to the caves. So much snow had fallen in such a way that it had been too great a weight for the mountain to hold. It had come down in a great avalanche, taking trees and boulders with it. Where the mouths of the caves would have been there was only a slanting white expanse of snow. It glittered from the sleet that had fallen the night before, making it a bright sheet of impassable ice. Even the trail they would have followed had been wiped away.

Sweetgrass Girl felt her legs weaken and she fell to her knees.

Rabbit Stick knelt beside her. "Kita," he said, "listen. There are openings into many of the caves from the top, up there." He pointed his chin up the ridge.

Sweetgrass Girl looked up at him. The certainty in his face made her stand up and look in that direction.

"But how can we find the right opening?" she said.

Rabbit Stick patted his small dog on its head. "Mikwe will go and look for us. He is so light that he will not break through the snow crust, and he is as sure footed as the little tree climbers."

Rabbit Stick went down on one knee and whispered into the small dog's ears. Mikwe stood up on his hind legs and placed one paw on the boy's shoulder and licked his face. Rabbit Stick stood.

"He is ready." He raised one hand and then held it outward. "Su-su gwilahwe!"

Mikwe ran up the slope, a small figure growing smaller and smaller until he reached the ridgetop where he was lost from sight.

There was a tightness in Sweetgrass Girl's throat that would not go away. She hoped but could not believe that the little dog would succeed. The Day Traveler seemed to be stuck in the sky and everything around her seemed as if it was blurred like the edges of a dream. Rabbit Stick took off one of his beaver robes and spread it in a spot sheltered from the light wind that was blowing, rattling the leaves of a beech tree just behind them.

"You can sit," he said. His voice was very kind.

Sweetgrass Girl sat. Moosis curled himself at back. She leaned her head forward and the weariness she had been fighting overcame her. She slept sitting, her arms and head on her knees.

"He is back!" Rabbit Stick's shout woke her. The little dog was running in circles at the boy's feet. "He is telling me that he has scented them. He will show us the way to get there."

The climb up the slope was not easy. The crust broke when they least expected it and their feet slipped often. But the two dogs were always there to help, the small one leading the way and the heavy one keeping close by them to steady their steps and help pull them out of the snow when they went in waist deep. At last they were on the ridgetop. There was almost no snow there for it had been scoured clean by a wind that cut like a flint blade.

Mikwe followed a trail that wove between the big stones, stopping at last where two big rocks leaned together and sheltered a deep hole between them just big enough for a man to enter.

"Be careful," Rabbit Stick said. "I have been here before—two summers ago. The hole goes straight down and it is a drop three times the height of a man."

Sweetgrass Girl leaned into the darkness and called. "My mother! My father!"

Her voice echoed back to her and then, more clearly than her own echoed words, she heard an answer from below.

"We are here." It was her mother's voice. "We lost our toboggan in the snow slide but escaped into this cave. Your father's leg is broken."

"We will help you," Sweetgrass Girl called back, excitement and relief in her voice. Already in her mind she could see them pulling her parents from the cave. She could see herself feeding them and wrapping them in the warm robes she had brought. She saw them making a toboggan from the bark of the trees on the slope below and using it to pull her injured father home safely.

Then she remembered. There was no way to reach them down there in the cave. The hole went straight down from the cave's ceiling. No one could climb out or climb in. She turned to look at Rabbit Stick, despair on her face. But Rabbit Stick was smiling and taking the beaver skin robes from his shoulders.

"Perhaps you are asking," he said, "how it is that I know it is exactly the height of three tall men from here to the floor of the cave?" He began to unsling the coils of braided basswood rope which he had wrapped about him, hidden until now under his robes. "It is because I used this rope two summers ago to lower myself down there."

Rabbit Stick smiled and braced himself. He knew that look on Sweetgrass Girl's face, having seen it appear just before she struck him with a stick when he had teased her too often.

But this time, to Rabbit Stick's surprise and delight, instead of hitting him, she wrapped her arms about him and pressed her face against his chest.

THE DANGEROUS STRANGER

AZEBAN SURPRISES CEDAR GIRL

It was a good morning to sleep. Molodokwskwasis or Cedar Girl pulled the deerskin robe back over her head for the third time and said it once more. "It is a good morning to sleep!" But this time she could not keep from giggling as she said it. She knew what was about to happen. Sure enough, with an indignant little growl, her dog Azeban grabbed the deerskin firmly between his teeth and tugged it down from her face.

Cedar Girl sat up, folded her arms in front of herself and tried to look stern. It was no use. Azeban sat there on his haunches, wagging his tail. His head was cocked to one side and the look on his face was so funny that she had to laugh.

"You bad dog," Cedar Girl said, failing to make her voice sound at all angry. "Do you not know I stayed up very late last night listening to stories being told? You were right there beside me. Everyone else is asleep still. Why should you and I get up?"

Azeban began to lift his front feet, one after another, to pat them on the good-smelling hemlock branches that were woven together to make a floor matting for the lodge of Cedar Girl's family. It made a sound almost like a drum. Whenever Azeban would do this, Cedar Girl's father Leaning Tree would always say that the wrong name had been chosen for her dog: "Instead of Raccoon, you should have named him Partridge!"

This morning as Azeban did his trick of drumming the floor with his paws, he struck so hard against one long branch that its butt end sprung free and stuck into the back of Cedar Girl's father where he and her mother were still sleeping on the other side of the wigwam. Leaning Tree's hand came slowly out from under the robes, reached up and back to dislodge the branch and then shoved it back down into place. As his hand disappeared back under the robes, Cedar Girl was sure she heard him murmur, "Partridge!"

Cedar Girl crawled to the door and pushed aside the skin that covered it. She looked outside, expecting to see the white hare. Every early morning since the leaves had fallen that hare had come to sit near their lodge. It was as if it was waiting to greet Cedar Girl, for it always sat until she looked through the lodge door. Then it looked at her and ran away. Today, though, the white hare was nowhere to be seen. That troubled Cedar Girl. Then, as Azeban drummed his paws again, she turned and dropped the door flap.

"Azeban," Cedar Girl said, "sit still."

Azeban stopped drumming his paws, but the look of excitement did not leave his eyes. If anything, his tail wagged back and forth even faster.

"My friend," Cedar Girl said, "I do not think even the squirrels are awake yet. Mikwe will wait for you."

At the sound of the word for squirrels, Azeban's ears pricked up, and he turned his head to look toward the door flap of their winter camp wigwam. Leaning Tree was known for being one of the best builders of such a camp, making its base of logs laid down to form a square which, when chinked with moss and lined with spruce boughs, kept out even the coldest winds. Even the snows would not drift in through the door, for the door was set up in the wall so that one had to climb up and over the height of three logs to get out.

Cedar Girl sighed. More, it seemed, than anything else in the world, Azeban loved to chase squirrels. Cedar Girl had explained to him that the white hare was her friend and so he paid no attention to it. Squirrels, though, were a different matter. Azeban knew each tree where a squirrel had escaped him in the past—for he never caught one—and would run to one after another of those trees each morning when he came bursting out of the lodge. Sometimes, in his excitement, he would even run past a squirrel sitting on a fallen branch to reach the base of the tree where that squirrel had escaped him the day before.

"Unnh-hunnh," she said. "Your enemies are out there and you think that you must protect us from them. But have you not heard the story that Stands-in-a-Hole told about long ago? The squirrels are no longer great monsters which threaten the people?"

"Mmmm-kwwwuuuuuh," Azeban whimpered. It was a sound he only made when he was begging to be allowed to go out and chase the squirrels. It sounded much like the word human people used for Mikwe, the squirrel. Cedar Girl shook her head.

"Hahh," she said. "It is no use for you to try to talk, Azeban! Did you not hear the story which Stands-in-a-Hole told us just last night? Long ago, you and your people could talk just as we human people talk. But the dogs were always talking too much. They talked so much they frightened the game away when they went hunting with their human friends. They talked so much that they kept the people awake at night. They bragged so much to all the other animals about what great hunters the human people were when the dog people helped them, that the other animals became afraid and started to move far away where they could not be found. It began to be hard for the people to find game when they went hunting and still the dogs would not stop talking. Finally, Gluskabe came to make things better. He told the dog people that they would no longer be able to talk with their voices as human people do."

Azeban lowered his head and looked up at Cedar Girl in a way that spoke to her more clearly than words. She knelt down and stroked her dog's wide forehead.

"My friend," she said, "you are good. And you can speak to me clearly with your eyes. Now I will stop teasing you. You can go and try to defeat your great enemy."

Cedar Girl pulled up the heavy skin robe that covered the door of their winter lodge; the brightness of the winter day, of the sunshine glittering on the white snow, came streaming in. Azeban stuck his head out next to hers, and, as soon as he did so, a red squirrel chirred from the top of a short broken tree a few paces from the lodge. Cedar Girl recognized that squirrel for it had only half a tail. It was not, though, because Azeban had almost caught it. During the last Moon of Falling Leaves, a big-shouldered hawk had swooped down behind that squirrel one day when it was engaged in what seemed to be its favorite pastime—sitting just out of reach and teasing the dog. The hawk and the squirrel had both tumbled to the ground.

It would have been that squirrel's last day had not Azeban leaped forward, barking as loudly as two dogs twice his size. The hawk had rolled to its back, lifting its feet up to protect itself from the dog with its talons. Azeban had stopped short, straddling the small, stunned squirrel as he continued to bark at the hawk until it righted itself, and, with great indignation, flapped back up into the sky. Carefully, Azeban had stepped backward and lain flat on the ground, chin on the earth, eyes on the little squirrel. Cedar Girl had held her breath as she watched, not certain what would happen next.

The little squirrel shook itself, then sat up and began to groom its fur. It was so close to Azeban that it surely could feel the dog's breath. Then it turned and bounded once, twice, three times, to the base of its favorite cedar where it leaped up and flattened itself against the trunk. Azeban's eyes followed the squirrel, but he still lay flat on the ground among the red and yellow leaves of the maples and the beeches. The squirrel looked around, then scooted behind the tree, appearing at last on a branch that hung over their clearing. It flicked its tail, which was noticeably shorter than it had been. Then, in a voice full of that familiar challenging tone the squirrel people use when they are scolding big ones watching from below, the red squirrel chirred and chirped. That was when Azeban leaped up, barked twice with excitement, ran to the base of the tree and began jumping up and down.

And so it was this morning. At the sound of his favorite enemy, Azeban leaped through the door and ran to the base of the tree. His claws scrabbled at the bark as if he thought he could actually climb up it. Somehow he even managed to hold on and pull himself a small way up the slanting trunk before sliding back down. That was why he had been given the name of Raccoon. More than any other dog Cedar Girl had known, Azeban wanted to climb into the trees. When a tree was slanted enough or there were enough big branches low to the ground, he would crawl up until, sometimes, he would get so high that Leaning Tree would have to climb up and bring him down before he fell and hurt himself.

"Azeban," Cedar Girl said, watching her words shape the white breath that came from her mouth, "do you want to chase the squirrel people or do you just want to join them?"

Azeban looked over at her from the place where he was now sitting at the base of the tree, waiting for the little squirrel to show itself again. The innocence in his eyes made her want to laugh again. Cedar Girl stepped out of the lodge and made her way along the well-used trail that led around the back through the brush toward the small, dark shelter made of boughs that everyone in her family visited each morning. Azeban did not follow her; he was too busy with his game.

It was not until Cedar Girl came out of the small lodge made of boughs that she realized something was wrong. At this time of the early morning, even in the winter, there were always birds along the brushy trail, singing their greetings to the new day. Since she was always the first to rise, she almost always surprised animals along that trail. She would see small deer or perhaps a porcupine or a white weasel, and for more than a moon now she had always seen the white hare. Whenever she went to the small lodge she would always see it sunning itself in front of its home in the briars. But this morning there had been no birds and the rabbit was still gone. That was strange. She looked more intently around her and her eyes caught sight of something that frightened her. It was the clear print of a man's foot in the snow, a fresh print which showed her by its shape that the one who wore that moccasin was not a member of her village.

Cedar Girl began to walk a little more quickly. The Only People had no human enemies. The days when they had been hunted by such great monsters as the giant bears and other terrible beings were long past. But to see the print of a stranger's foot made her worry that this might be someone from a distant people, someone whose ways were different. There were stories of men who could not speak like real human beings and who would sneak close to the village to try to kidnap a young woman to bring her back to his own people and make her his wife.

The sound of her own breathing became very loud in Cedar Girl's ears. It seemed as if it would be hard to hear anything else. She did hear something. She heard the sound of the brush rustling and catching on someone's deerskin clothing behind her and the thump of feet. When she turned, she caught only a quick glimpse of the tattooed face of the man who threw a skin over her head to muffle her cries. Then she could see nothing but darkness. Though she struggled and kicked her feet, she was picked up and carried away.

The one who had taken her was strong, for he threw her over his shoulder as easily as if she had been a rabbit. He carried her for a long time. The skin was so tight around her head that she could hardly breathe. When she was put down at last and the skin was pulled from over her head, she felt weak and sick. She looked around. They had come a long way, at least four looks from the village. They were on the rocky hilltop at the edge of her father's hunting territory. The big pine tree behind the man who had taken her was one she had often seen. Its low big branches hung over them and it was marked with the sign of a slanted tree, her father's sign.

Cedar Girl looked at that sign, not wanting to look at the man who had stolen her. There was sickness in her heart at the thought that this man was going to take her far away from this familiar place. She turned her eyes to that man, whose face was hard to see, for it was painted black from chin to eyes and a curling tattoo shaped like a snake went across his forehead. The man who stood below the overhanging branches of the pine was even bigger than her father and dressed strangely. She lifted her hand to her mouth to keep from crying. Hung from his belt, blood dried on its nose, was the dead body of the white hare that had greeted her so many mornings.

The man was gesturing to her. She understood. He wanted her to hold out her hands so that he could bind them with the sinew cord which he held in his hands. Then he could pull her along with no worry about her running away.

Cedar Girl saw something moving on the wide pine branch that slanted up from the ground above the man. It was too large to be a squirrel, too small to be a long-tail cat, just a bit larger than a big raccoon. The tattooed man, seeing her look up, turned himself halfway around and lifted his chin. Just as he did so, with a low growl, Azeban leaped down from the tree, striking the big man in his chest with a loud thump! The man fell back heavily among the stones with the dog on top of him. Azeban leaped free, snarling, ready to attack. But the man did not get up.

Cedar Girl approached the man cautiously. His eyes were closed and he was breathing, but he had struck the rocks hard and could not move. She reached down, pulled the hare free from his belt and began to run, Azeban running beside her. They ran hard, down the hill and along the trail that wound through the ridges. At the top of the next hill, Cedar Girl looked back. No one followed them. She still held the body of the white hare. It was too heavy to carry further, but she did not want to allow the stranger to find the body of that animal which had greeted her every dawn. Carefully she placed the body of the white hare in a deep crack between the rock and piled stones over it. Then she stood and began to run again.

By the time they came to the edge of the village, the Day Traveler was only two hands above the horizon and the shadows of evening beginning to arrive. Though Azeban had circled back many times to be sure they were not pursued, Cedar Girl was certain now that the tattooed man had not followed them. She was sure that, wherever he had come from, he was well on his way back toward his own home. Perhaps he had learned his lesson and would not come again to bother her people. Her own lodge in sight, Cedar Girl knelt down and took Azeban's wide head in her hands.

"My friend," she said, "never again will I tease you for wanting to climb trees like the squirrel people. You are truly my Azeban, my raccoon dog."

GLOSSARY

Abenaki: Dawn Land; also refers to the People of the Dawn Land
Alnobak: Human Beings
Awasos: Bear
Awasosis: Little Bear
Awasosqua: Bear Woman
Azeban: Raccoon
Gluskabe: The Talker or The Storyteller; the trickster and transformer hero of Abenaki stories
Ki: Land or earth
Kina: Look here
Kita: Listen
Ktsi: Big or great
Ktsindatlogit: Great Storyteller
Ktsi Nwaskw: Great Spirit; the Creator
Kwanitewk: Long River; the Connecticut River
Kwaniwibid: Long Tooth
Mahtegwas: Rabbit
Mahtegwasabazi: Rabbit Stick
Mikwe: Squirrel
Molodowkskwasis: Cedar Girl
Moos: Moose
Moosis: Little Moose
Moskwaso: Muskrat
Naboomsawinno: Keeps-Following-the-Trail
Nanatasis: Hummingbird
Nda: No
Ndakinna: Our Land; roughly the area now known as Vermont and northern New England
Nidobak: My friends
Nolka: Deer
Oleohneh: Thank you (also may be spelled Wliwini)
Petonbok: Waters in Between, Lake Champlain
Soksemo: Good Nose
Su su gwilawe: Go and fetch it
Tomakwa: Beaver
Unh-honh!: Yes (also may be spelled *Onh-honh*)
Waban: Dawn
Wabanki: Dawn Land
Walimogwkilskwasis: Sweetgrass Girl